Published in 2024 by Ruby Tuesday Books Ltd.

Copyright © 2024 Ruby Tuesday Books Ltd.

All rights reserved. No part of this publication may be reproduced in whole or in part, stored in any retrieval system, or transmitted in any form or by any means, electronic, mechanical, photocopying, recording, or otherwise, without written permission from the publisher.

Editor: Mark J. Sachner
Design: Emma Randall
Production: John Lingham

Photo credits:
Alamy: 17 (Biosphoto); Ruby Tuesday Books: 7, 9T, 20B, 21; Shutterstock: Cover, 4 (nelea33/Yeti Studio/Werner Muenzker/bergamont/Hortimages/Spalnic/topseller/Harun Ozmen/Golden Shark 2), 5T (Hakan Tanak), 5B (kurbanov), 6T (Boltenkoff), 6B (Pavel L Photo and Video), 7 (Africa Studio/montego), 8T (Kazakova Maryia), 8B (makeevadecor/Mironmax Studio), 9B (Kate Scott), 10 (Khadi Ganiev/Blue Ring Media), 11T (M. Unal Ozmen/New Africa/Madlen), 11C (pundapanda), 11B (Sstudi), 12–13 (Khadi Ganiev/Madlen/ajt/Diana Taliun/Paul Maguire), 14–15 (Khadi Ganiev/Anton Starikov/AlenKadr/Jiang Hongyan/Lopatin Anton), 16T (Riepina Vladyslava), 16B (Kazakova Maryia), 18T (O Lypa), 18B (PrysMichael/Natalia Kokhanova), 19T (Khadi Ganiev), 19C (Brilliance Stock), 19B (Valentyn Volkov), 20T (Rolandas Grigaitis/Mikhail Martynov), 21 (Picture Partners/mayakova/Vita Sorokina/Keith Homan/Pixel-Shot/AmaPhoto/vitals), 22 (Anastasiia Voloshka/Alexander Raths), 23 (Smeerjewegproducties/malshkoff/schankz).

British Library Cataloguing in Publication Data (CIP) is available for this title.

ISBN 978-1-78856-344-4

Printed in Poland by L&C Printing Group

www.rubytuesdaybooks.com

Contents

The Main Ingredient 4
It All Begins with Seeds 6
The Shoots Grow 8
It's Time for Transplanting 10
Growing Bigger 12
Caring for Your Plant 14
Bumblebee Teamwork 16
Your Crop Is Ready! 18
Let's Make Ketchup! 20
Glossary 22
Index 24

The Main Ingredient

Sweet, juicy tomatoes taste good in salads and sandwiches. We also use them as an **ingredient** when we make pizzas, chilli con carne, pasta sauce and KETCHUP.

People grow tomatoes all over the world. There are about 10,000 different types.

Tomatoes came from South America. They grew as wild plants in countries such as Peru and Chile.

The Aztec people of Mexico were possibly the first to grow tomatoes as a food crop around 1500 years ago. The Aztec word for "tomato" was tomatl. It meant plump fruit with a belly button!

Tomato plant

Tomatoes growing on a farm

Greenhouse

Today, farmers around the world grow tomatoes in huge **greenhouses**. People grow them in gardens, too.

In this book, you will learn how to grow a tomato plant, AND turn its tomatoes into delicious ketchup!

It All Begins with Seeds

The tiny pips inside a tomato are **seeds**. It's possible to grow a tomato plant from these seeds. However, most gardeners buy tomato seeds in a packet from a garden centre, shop or online.

Tomato seeds

Buying seeds at a garden centre

YOU WILL NEED:

- A packet of tomato seeds
- 6 small flowerpots or yogurt cartons (with holes made in the bottom)
- A small trowel
- Potting compost
- A watering can
- 6 saucers (or a tray large enough for 6 pots or cartons)
- A bucket-sized container with holes in the bottom
- A bamboo cane or stick about 1.25 m long
- String or wool
- Scissors
- Shop-bought tomato feed (or see page 15 for homemade recipe)

Seeds that we buy in packets have been specially grown and cared for to make sure they grow into healthy plants.

1. Choose and buy a packet of tomato seeds from a garden centre or shop. You can also ask an adult to help you buy seeds online.

2. Put potting compost into your small pot or carton until it is about three-quarters full.

3. Place six tomato seeds on the surface of the soil.

Always plant more seeds than you need, just in case some seeds don't grow.

4. Next, cover the seeds with about 2 cm of soil. Gently water the pot so the soil is **moist**.

5. Place the pot on a saucer and stand on a sunny windowsill.

The saucer will catch any water that trickles from the pot.

The Shoots Grow

Under the warm soil, your tiny tomato seeds will grow thin, thread-like **roots**. Next, each seed will produce a **shoot**. The shoot pushes up through the soil to get to the light.

Seed

Root

On a warm, sunny windowsill, it should take about two weeks for the shoots to appear.

A tiny shoot uncurling.

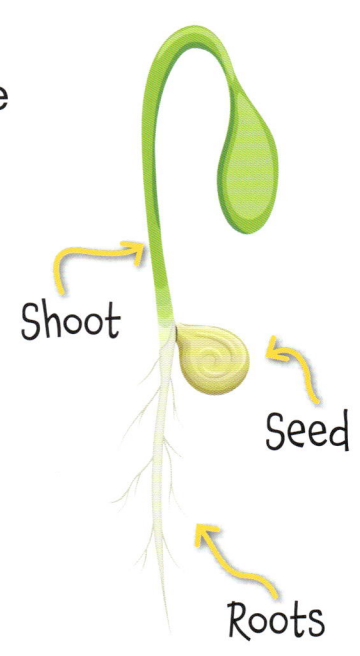
Shoot
Seed
Roots

Seed leaves

1. After one or two days, a pair of tiny leaves will unfurl from each shoot. These leaves are called seed leaves.

A shoot gets energy and nutrients from its seed leaves.

2. Gently touch the soil in the pot each day with your finger. If the soil feels dry, water the shoots.

3. After a few days, you will notice a second pair of leaves emerging from the tip of each shoot.

The second pair of leaves are known as proper tomato plant leaves.

Seed leaf

Tomato plant leaf

4. Make sure the little plants, or seedlings, are in a warm, sunny position. Water them to keep the soil moist.

It's Time for Transplanting

Your tomato seedlings will grow bigger each day and produce more leaves. All this growth uses lots of energy.

The plants' leaves make their own sugary food for energy. They do this using water, **carbon dioxide** gas and sunshine.

Leaves:
The leaves take in carbon dioxide gas from the air. They do this through stomata, which are microscopic holes that open and close.

Open stomata

Closed stomata

Roots:
The plant's roots suck up water from the soil.

When leaves make food, the process is called **photosynthesis**.

When your seedlings are about 8 cm tall, it is time to transplant each one into its own pot.

1 Half fill six small flowerpots with soil. You can also use empty yogurt cartons or other recycled containers.

Make sure a recycled container has holes in the bottom.

2 Tap the pot of seedlings to loosen the soil. Then gently tip the pot onto its side so the plants and soil slide out. Carefully separate the little plants.

Each seedling should have its own clump of soil-covered roots.

Do not touch a seedling's stem – you might crush it! Hold the leaves as the seedling can grow more leaves but it can't grow a new stem.

3 Take hold of a seedling by a leaf and lower it into a pot of soil. Add more soil, nearly to the top of the pot. Then gently press down the soil around the plant's stem.

Repeat for the other five seedlings.

4 Water the plants, place them on saucers or a tray, and put them somewhere warm and sunny.

Growing Bigger

We transplant seedlings into their own pots so they are not competing for space, water and nutrients in the soil.

1 Every day, check if your plants need watering. Tomato plants like LOTS of water. However, the soil should be moist, but not soggy.

The plants' stems will grow taller and thicker. More leaves and roots will grow, too.

2 When the plants are 20 cm tall, they are ready to be transplanted to their final growing place. You can take care of all six plants or choose one and give the others to friends.

3 For a single tomato plant, fill a bucket-sized flowerpot or other container with potting compost. Then, using your spade, scoop out a hole the same size as the pot that contains your tomato plant.

4 Take the pot containing your tomato plant and carefully tap and tip it onto its side so the plant slides out. Put the plant into the hole inside the new, bigger container. Scoop the soil around the plant's stem and gently press it down.

Always wash your hands with soap and warm water after working with soil.

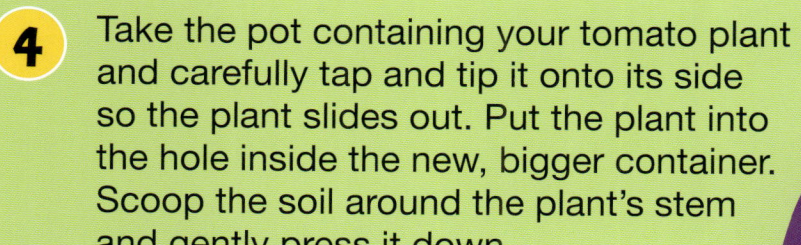

5 Continue to check and water your plant each day to keep the soil moist. The plant can now be placed outdoors in a sunny spot.

You might spot new shoots growing where the plant's leaves join the main stem. These side shoots will take energy away from the plant's flowers and tomato fruits, so carefully break them off.

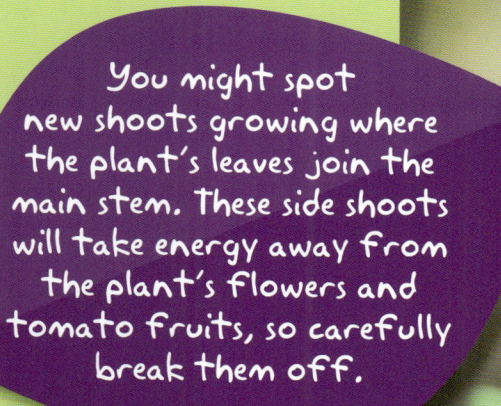

Side shoot

Main stem

Leaf

Caring for Your Plant

Your tomato plant will grow taller and taller and produce more leaves.

1 Once your plant is 45 cm tall, push a long stick into the pot alongside the plant.

2 Carefully tie the plant's main stem to the stick to stop it falling over. You can use string, wool or special plant ties.

3 Your plant will soon be ready to grow flowers. Once this happens, you will need to add extra nutrients to the soil. You can buy tomato plant feed from shops or get ready to feed your plant by making your own.

Tomato plant feed is mixed with water. Follow the instructions on the bottle or packet.

4. To make tomato feed, put four banana skins into a large jar with a lid. Fill the jar with water and tighten the lid.

5. Leave the jar in a warm place for about two weeks.

Yellow flowers

6. When yellow flowers start to appear, pour the jar of banana peel tomato feed into the plant's container. You can add the mushy peels to the soil, too. This will add lots of nutrients to the soil.

Your plant will take in nutrients from the feed with its roots when it sucks up water. The extra nutrients will help the plant grow tomatoes.

Bumblebee Teamwork

When your tomato plant is about two to three months old, it will grow yellow flowers.

Now **fertilisation** must take place so each flower can produce a tomato.

Tomato flowers

Tomato Flower Fertilisation

- The flower's anthers produce dusty **pollen**.

- The pollen must land on the flower's stigma.

- Each pollen grain sends a tiny tube down the style into the ovary.

- The pollen tubes enter the tiny ovules. Now the ovules are fertilised and will become tomato seeds.

- The ovary starts to swell up around the seeds and becomes a tomato.

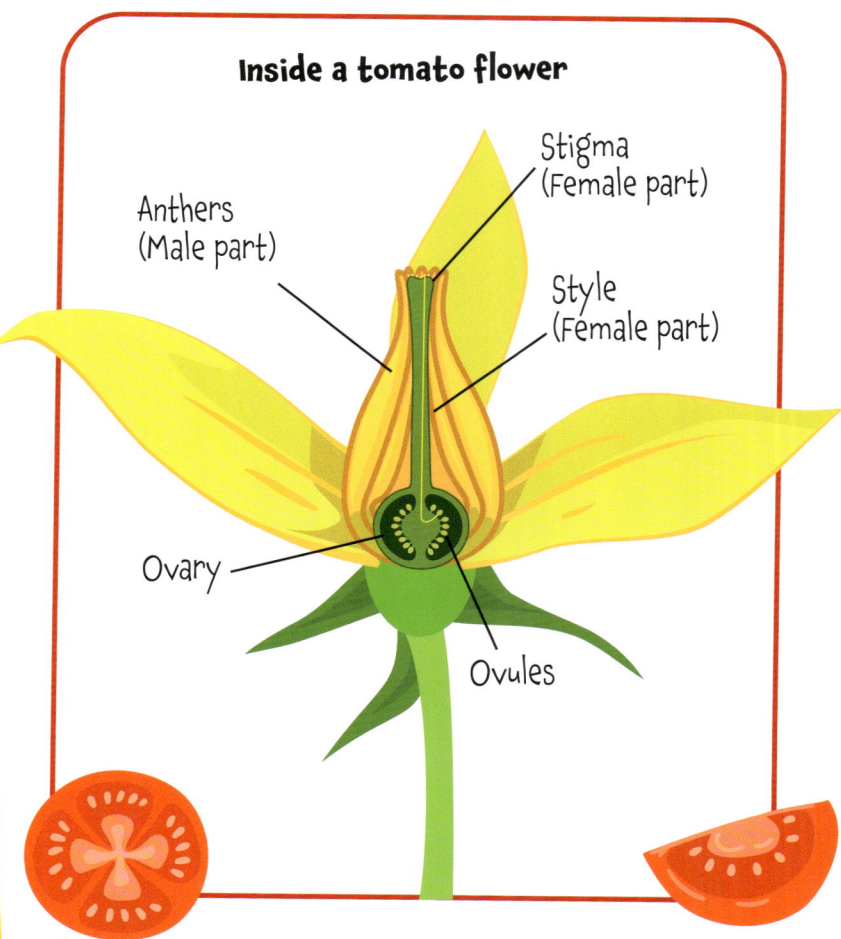
Inside a tomato flower

How will your tomato flowers be fertilised? You will need help from bumblebees!

1 A bumblebee visits a tomato flower to collect sweet nectar and pollen.

2 The bee vibrates her body, which shakes loose pollen from the flower's anthers. Some pollen lands on the flower's stigma.

Tomato flower

Pollen

Bumblebee vibrating

Pollen basket

3 Some pollen sticks to the bee's furry body. She uses her legs to comb pollen from her fur into her pollen baskets and takes it back to her nest.

4 Some pollen stays stuck to the bee's body. When she visits another flower, this pollen may brush off onto the stigma of that flower.

The way that bumblebees help **pollinate** tomato plants is called buzz pollination.

Your Crop Is Ready!

Once your tomato plant's flowers have been fertilised, they will start to die. What happens next?

1

New tomato

Dead flower

Keep checking the dying flowers. You will soon see a tiny green tomato growing from each one.

2 Water your plant every morning with 1 litre of water. If the weather is hot and sunny, give the plant the same amount of water in the evening.

Make sure your plant's soil is always moist and doesn't get dry.

3 If you see any leaves that look yellow, shrivelled or unhealthy, cut them off. If any fruits look damaged or unhealthy, carefully cut them off, too.

Yellow leaf   Unhealthy tomato fruit

Green tomato

Ripe tomato

Roots

Water and nutrients

4 Your green tomatoes will grow bigger and bigger.

5 Feed your tomato plant once every two weeks with homemade tomato feed or shop-bought feed mixed with water. Pour the feed into the pot around the plant's stem.

6 Each tomato will turn from green to red.

Ripe tomato

7 Once a tomato is fully red, it is **ripe**. Gently pull it from the plant.

8 Your plant will keep on growing flowers and tomatoes for up to three months.

We think of a tomato as a vegetable, but it is actually a fruit!

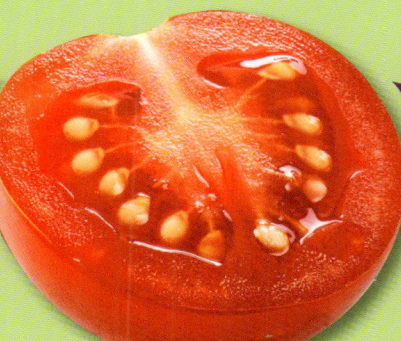

A fruit is the part of a plant where its seeds grow.

Let's Make Ketchup!

If you love to eat ketchup, this recipe will show you how to make it with your homegrown tomatoes.

You can use large tomatoes or small cherry tomatoes.

BE SAFE!
Be sure that an adult is there to help you when you are using a knife and the hob.

YOU WILL NEED:

Ingredients
- 4 or 5 large tomatoes or a heaped cup of cherry tomatoes (washed)
- 4 teaspoons apple cider vinegar
- 2 teaspoons Worcestershire sauce
- 4 teaspoons light brown sugar
- 3 teaspoons salt
- 2 teaspoons paprika

(The ingredients will make about half a cup of ketchup.)

Equipment
- Measuring cups and spoons
- A knife and chopping board
- A small saucepan
- A wooden spoon
- A trivet
- A hand blender or food processor
- An airtight container with a lid

1 If you are cooking with big tomatoes, carefully cut each one in half and cut out and throw away the core. Next, chop the tomatoes into small pieces.

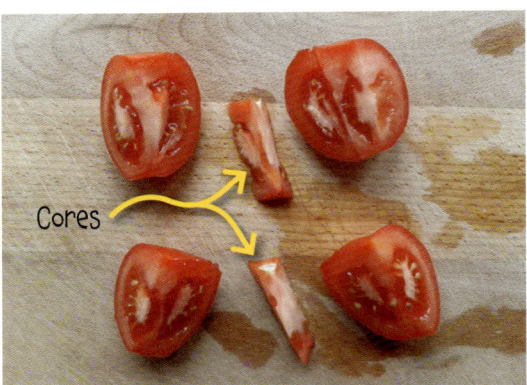

Cores

If you are using cherry tomatoes, chop each one into quarters.

You will need enough chopped tomatoes to fill a measuring cup or coffee mug.

2 Put the tomatoes into a small saucepan. Add the apple cider vinegar, Worcestershire sauce, sugar, salt and paprika.

Light brown sugar  Salt   Paprika

3 Put the pan onto the hob on a high heat. Keep stirring the mixture until it is bubbling and boiling.

4 Once the mixture is boiling, turn down the heat to medium. Stir the mixture with a wooden spoon about every 30 seconds. Cook for 10 minutes.

5 Remove the saucepan from the hob and place on a trivet to cool.

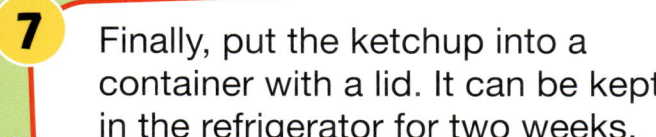

6 When the mixture is cool, use a hand blender or food processor to make the ketchup smooth.

7 Finally, put the ketchup into a container with a lid. It can be kept in the refrigerator for two weeks.

Keep on picking and eating your tomatoes, and enjoy your homegrown ketchup!

Glossary

carbon dioxide
A colourless gas in the air that plants use to make food.

fertilisation
In some plants, the process in which a pollen grain enters an ovule inside a flower. This makes the ovule ready to grow into a seed.

greenhouse
A large building made mostly of glass or clear plastic that is warm inside. It is used for growing fruits, vegetables and other plants.

ingredient
A food or substance that is used to make a particular dish or meal.

moist
Slightly wet.

nutrients
Substances needed by a plant or animal to help it live and grow. For example, potassium is a nutrient that helps tomato plants grow healthy fruit.

photosynthesis
The process by which plants make food in their leaves using water, carbon dioxide and sunshine.

pollen
A dust that is made by flowers. It is needed for fertilisation and making seeds.

pollinate
To transfer pollen from one flower to another.

ripe
Ready to be picked.

roots
Underground parts of a plant that take in water and nutrients from the soil.

seed
A tiny part of a plant that contains all the material needed to grow a new plant.

shoot
A new part of a plant. Shoots grow from seeds in soil. Also, when a new twig, leaf or flower grows from a stem, it is called a shoot.

Index

A
Aztecs, the 5

B
bumblebees 16–17

F
fertilisation 16–17, 18
flowers 14–15, 16–17, 18–19

K
ketchup 4–5, 20–21

L
leaves 9, 10–11, 12–13, 14, 18

P
photosynthesis 10
planting seeds 6–7
pollen 16–17

R
roots 8, 10–11, 12, 15, 19

S
seedlings 9, 10–11, 12
seeds 6–7, 8–9, 16, 19
shoots 8–9, 13
side shoots 13

T
tomato farms 5
tomato plant feed 9, 14–15, 19
transplanting 10–11, 12–13

W
watering 7, 9, 11, 12–13, 15, 18–19